Pinpoint Math

Student Booklet
Level D

Volume 5
Geometry and Measurement

Photo Credits
©iStock International Inc., cover.

Acknowledgements
Content Consultant:

Linda Proudfit, Ph.D.

After earning a B.A. and M.A in Mathematics from the University of Northern Iowa, Linda Proudfit taught junior- and senior-high mathematics in Iowa. Following this, she earned a Ph.D. in Mathematics Education from Indiana University. She currently is Coordinator of Elementary Education and Professor of Mathematics Education at Governors State University in University Park, IL.

Dr. Proudfit has made numerous presentations at professional meetings at the local, state, and national levels. Her main research interests are problem solving and algebraic thinking.

www.WrightGroup.com

Copyright © 2009 by Wright Group/McGraw-Hill.

All rights reserved. Except as permitted under the United States Copyright Act, no part of this publication may be reproduced or distributed in any form or by any means, or stored in a database or retrieval system, without the prior written permission from the publisher, unless otherwise indicated.

Printed in USA.

Send all inquiries to:
Wright Group/McGraw-Hill
P.O. Box 812960
Chicago, IL 60681

ISBN 978-1-40-4567979
MHID 1-40-4567976

5 6 7 8 9 10 RHR 13 12 11

Contents

Tutorial Chart vii

Volume 5 Geometry and Measurement

Topic 13: Basic Geometric Figures
Topic 13 Introduction...1
Lesson 13-1 Draw Angles and Lines.......................................2–4
Lesson 13-2 Types of Polygons 5–7
Lesson 13-3 Triangles..8–10
Lesson 13-4 Quadrilaterals...11–13
Topic 13 Summary..14
Topic 13 Mixed Review...15

Topic 14 Standard Measurement
Topic 14 Introduction 16
Lesson 14-1 Basic Units of Length......................................17–19
Lesson 14-2 Basic Units of Weight......................................20–22
Lesson 14-3 Basic Units of Liquid Capacity.............................23–25
Topic 14 Summary..26
Topic 14 Mixed Review...27

Topic 15 Metric Measurement

Topic 15 Introduction..28
Lesson 15-1 Basic Metric Prefixes 29–31
Lesson 15-2 Basic Metric Units of Length....................32–34
Lesson 15-3 Basic Metric Units of Mass.......................35–37
Lesson 15-4 Basic Metric Units of Capacity.................38–40
Topic 15 Summary..41
Topic 15 Mixed Review 42

Topic 16 Measure Geometric Figures

Topic 16 Introduction 43
Lesson 6-1 Length...44–46
Lesson 16-2 Perimeter...47–49
Lesson 16-3 Area 50–52
Lesson 16-4 Area of Rectangles.................................53–55
Lesson 16-5 Volume 56–58
Topic 16 Summary 59
Topic 16 Mixed Review 60

Glossary..61

Word Bank 63

Index 65

Objectives

Volume 5: Geometry and Measurement
Topic 13 Basic Geometric Figures

Lesson	Objective	Pages
Topic 13 Introduction	13.1 Draw, measure, and classify different types of angles and lines. 13.2 Define *polygon* and classify the different types of polygons. 13.3 Explore, compare, and classify different types of triangles. 13.4 Explore, compare, and classify different types of quadrilaterals.	1
Lesson 13-1 Draw Angles and Lines	13.1 Draw, measure, and classify different types of angles and lines.	2–4
Lesson 13-2 Types of Polygons	13.2 Define *polygon* and classify the different types of polygons.	5–7
Lesson 13-3 Triangles	13.3 Explore, compare, and classify different types of triangles.	8–10
Lesson 13-4 Quadrilaterals	13.4 Explore, compare, and classify different types of quadrilaterals.	11–13
Topic 13 Summary	Review basic geometric figures.	14
Topic 13 Mixed Review	Maintain concepts and skills.	15

Topic 14 Standard Measurement

Lesson	Objective	Pages
Topic 14 Introduction	14.1 Explore the basic units of length in the U.S. customary system. 14.2 Explore the basic U.S. customary units of weight. 14.3 Explore the basic U.S. customary units of liquid capacity.	16
Lesson 14-1 Basic Units of Length	14.1 Explore the basic units of length in the U.S. customary system.	17–19
Lesson 14-2 Basic Units of Weight	14.2 Explore the basic U.S. customary units of weight.	20–22
Lesson 14-3 Basic Units of Liquid Capacity	14.3 Explore the basic U.S. customary units of liquid capacity.	23–25
Topic 14 Summary	Review standard measurements.	26
Topic 14 Mixed Review	Maintain concepts and skills.	27

Topic 15 Metric Measurement

Lesson	Objective	Pages
Topic 15 Introduction	15.1 Explore the basic metric prefixes and what they mean. 15.2 Explore the basic metric units of length. 15.3 Explore the basic metric units of mass. 15.4 Explore of basic metric units of liquid capacity.	28
Lesson 15-1 Basic Metric Prefixes	15.1 Explore the basic metric prefixes.	29-31
Lesson 15-2 Basic Metric Units of Length	15.2 Explore the basic metric units of length.	32-34
Lesson 15-3 Basic Metric Units of Mass	15.3 Explore the basic metric units of mass.	35-37
Lesson 15-4 Basic Metric Units of Capacity	15.4 Explore of basic metric units of liquid capacity.	38-40
Topic 15 Summary	Review metric measurement.	41
Topic 15 Mixed Review	Maintain concepts and skills.	42

Topic 16 Measure Geometric Figures

Lesson	Objective	Pages
Topic 16 Introduction	16.1 Measure the length of an object to the nearest inch or centimeter. 16.2 Find the perimeter of a polygon with integer sides. 16.3 Estimate or determine the area of figures by covering them with squares. 16.5 Estimate or determine the volume of solid figures by counting the number of cubes that would fill them.	43
Lesson 16-1 Length	16.1 Measure the length of an object to the nearest inch or centimeter.	44–46
Lesson 16-2 Perimeter	16.2 Find the perimeter of a polygon with integer sides.	47–49
Lesson 16-3 Area	16.3 Estimate or determine the area of figures by covering them with squares.	50–52
Lesson 16-4 Area of Rectangles	16.4 Measure the area of rectangular shapes by using appropriate units.	53–55
Lesson 16-5 Volume	16.5 Estimate or determine the volume of solid figures by counting the number of cubes that would fill them.	56–58
Topic 16 Summary	Review measuring geometric figures.	59
Topic 16 Mixed Review	Maintain concepts and skills.	60

Tutorial Guide

Each of the standards listed below has at least one animated tutorial for students to use with the lesson that matches the objective. If you are using the electronic components of *Pinpoint Math,* you will find a complete listing of Tutorial codes and titles when you access them either online or via CD-ROM.

Level D

Standards by topic	Tutorial codes
Volume 5 Geometry and Measurement	
Topic 13 Basic Geometric Figures	
13.1 Draw, measure, and classify different types of angles and lines.	13a Measuring an Angle
13.1 Draw, measure, and classify different types of angles and lines.	13b Drawing an Angle
13.1 Draw, measure, and classify different types of angles and lines.	13c Identifying and Drawing Parallel Lines
13.1 Draw, measure, and classify different types of angles and lines.	13d Identifying and Drawing Perpendicular Lines
13.2 Define *polygon* and classify the different types of polygons.	13e Classifying Polygons
13.3 Explore, compare, and classify different types of triangles.	13f Sorting and Classifying Triangles
13.3 Explore, compare, and classify different types of triangles.	13g Finding Angle Measures in Triangles
13.4 Explore, compare, and classify different types of quadrilaterals.	13h Finding Angle Measures in Quadrilaterals
Topic 14 Standard Measurement	
14.1 Explore the basic units of length in the U.S. customary system.	14a Converting Units of Length
14.1 Explore the basic units of length in the U.S. customary system.	14b Measuring the Sides of a Box
14.1 Explore the basic units of length in the U.S. customary system.	14c Measuring Length to the Nearest Unit
14.3 Explore the basic U.S. customary units of liquid capacity.	14c Converting Units of Capacity
Topic 15 Metric Measurement	
15.1 Explore the basic metric prefixes and what they mean.	15a Using the Metric System to Measure Length
15.2 Explore the basic metric units of length.	15b Using the Metric System to Measure Mass
15.3 Explore the basic metric units of mass.	15b Using the Metric System to Measure Mass
Topic 16 Measure Geometric Figures	
16.1 Measure the length of an object to the nearest inch or centimeter.	16a Measuring the Sides of a Box
16.1 Measure the length of an object to the nearest inch or centimeter.	16b Measuring Length to the Nearest Unit
16.2 Find the perimeter of a polygon with integer sides.	16c Finding Perimeter
16.3 Estimate or determine the area of figures by covering them with squares.	16d Finding Area
16.5 Estimate or determine the volume of solid figures by counting the number of cubes that would fill them.	16d Finding Volume

Topic 13: Basic Geometric Figures

Topic Introduction

Complete with teacher help if needed.

1. Consider triangle *XYZ*.

   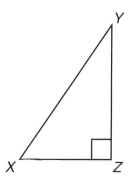

   a. Angle _____ is a right angle.

   b. Angles *X* and *Y* are called _____ angles.

   Objective 13.1: Draw, measure, and classify different types of angles and lines.

2. Consider triangle *ABC* and quadrilateral *DEFG*.

   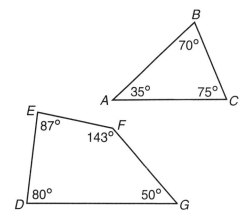

   a. The sum of the angles in the triangle is _____.

   b. The sum of the angles in the quadrilateral is _____.

   Objective 13.3: Explore, compare, and classify different types of triangles.
   Objective 13.4: Explore, compare, and classify different types of quadrilaterals.

3. Give the angle measures.

   a. A straight angle is _____.

   b. A right angle is _____.

   Objective 13.1: Draw, measure, and classify different types of angles and lines.

4. Classify these polygons.

   a. What polygon has 4 equal sides and 4 right angles? _____

   b. What polygon has three sides? _____

   c. What is the name of an eight-sided polygon? _____

   Objective 13.2: Define polygon and classify the different types of polygons.

Volume 5 — Level D

Lesson 13-1 — Draw Angles and Lines

Model It

Words to Know

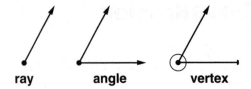
ray   angle   vertex

Activity 1

Measure the angle.

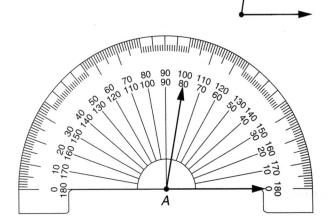

Read where the other ray crosses the inside scale. The angle measures 80°.

Practice 1

Measure the angle.

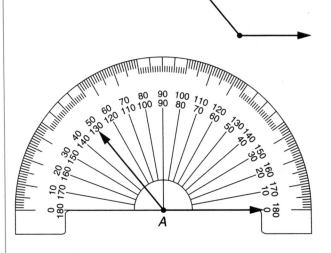

The measure angle. _____.

Activity 2

Draw an angle whose measure is 45°. Draw one ray. Align 0° on a protractor with that side. Mark a dot at 45° on the inside scale. Connect the dot to the endpoint of the ray.

Practice 2

Draw an angle whose measure is 90°.

On Your Own

Draw a 125° angle.

Write About It

How do you measure an angle that opens to the right?

Objective 13.1: Draw, measure, and classify different types of angles and lines.

Lesson 13-1 — **Draw Angles and Lines**

Understand It — B

Words to Know

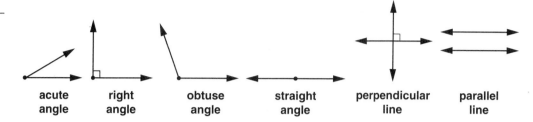

acute angle · right angle · obtuse angle · straight angle · perpendicular line · parallel line

---

Example 1

Measure each angle. Classify each angle.

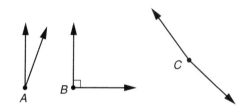

Angle A measures 20°, so it is **acute**.
Angle B measures 90°, so it is **right**.
Angle C measures 170°, so it is **obtuse**.

Practice 1

Measure. Classify each angle.

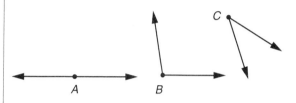

Angle A is a _____ angle.

Angle B is an _____ angle.

Angle C is an _____ angle.

Example 2

Name a pair of perpendicular lines.

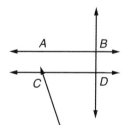

Lines $\overleftrightarrow{AB}$ and $\overleftrightarrow{BD}$ are perpendicular because they cross to form right angles.

Practice 2

Name a pair of perpendicular lines.

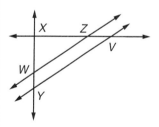

Lines _____ intersect because they cross to form _____.

On Your Own

Draw a pair of parallel lines.

Write About It

Describe something in real life that resembles parallel lines.

Objective 13.1: Draw, measure, and classify different types of angles and lines.

Lesson 13-1 **Draw Angles and Lines**

Try It

1. Find the measure of the angle.

_____

2. Draw an angle with a measure of 150°.

3. Which pair of lines is perpendicular? Circle the letter of the correct answer.

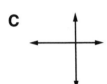

4. Which is a straight angle? Circle the letter of the correct answer.

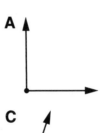

5. What is the measure of the angle? Circle the letter of the correct answer.

A 45°   **B** 70°

C 90°   **D** 110°

6. What kind of angle is this? Circle the letter of the correct answer.

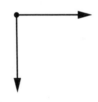

A acute   **B** right

C obtuse   **D** straight

7. Name a pair of intersecting lines in the diagram below.

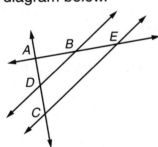

8. Explain how to use a protractor to draw a 120° angle. Then draw the angle.

Objective 13.1: Draw, measure, and classify different types of angles and lines.

Lesson 13-2 — Types of Polygons

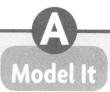

Words to Know A **polygon** is a closed figure made up of 3 or more line segments that meet but do not cross. The line segments are called **sides**. A **vertex** is the point where two sides of a polygon meet.

Activity 1

How many sides and vertices does a **triangle** have?

Model a triangle.

A triangle has 3 sides and 3 vertices.

Practice 1

How many sides and vertices does a **hexagon** have? Model a hexagon.

A hexagon has _____ sides and vertices.

Activity 2

How many sides and vertices does a **quadrilateral** have?

Model a quadrilateral.

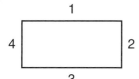

A quadrilateral has 4 sides and 4 vertices.

Practice 2

How many sides and vertices does a **pentagon** have? Model a pentagon.

A pentagon has _____ sides and vertices.

On Your Own

Do you notice a pattern among polygons number of sides and number of vertices?

_____

_____

Write About It

Explain why the figure below is not a polygon.

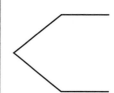

Objective 13.2: Define polygons and classify different types of polygons.

Lesson 13-2: Types of Polygons

Understand It — B

Words to Know: **Tri-** means 3. **Quad-** means 4. **Pent-** means 5. **Hex-** means 6. **Hept-** means 7. **Oct-** means 8. **Dec-** means 10.

Example 1

Classify the polygon.

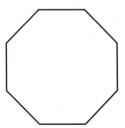

The polygon has 8 sides and 8 vertices. The polygon is an **octagon**.

Practice 1

Classify the polygon.

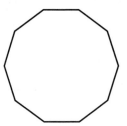

The polygon has _____ sides and vertices. The polygon is a _____.

Example 2

Classify the polygon.

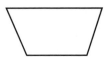

The polygon has 4 sides and 4 vertices.

The polygon is a **quadrilateral**.

Practice 2

Classify the polygon.

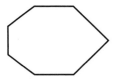

The polygon has _____ sides and vertices. The polygon is a _____.

On Your Own

Classify the polygon.

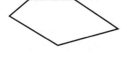

_____

Write About It

Explain why each figure is a quadrilateral.

  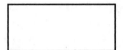

Objective 13.2: Define polygons and classify different types of polygons.

Lesson 13-2   **Types of Polygons**

1. How many sides does each shape have?

   a. triangle _____

   b. heptagon _____

   c. quadrilateral _____

2. Draw two different quadrilaterals. Then circle the vertices.

3. Which figure is **not** a triangle? Circle the letter of the correct answer.

   A   B

   C   D

4. What is the name of the polygon? Circle the letter of the correct answer.

   A pentagon   **B** hexagon

   C heptagon   **D** octagon

5. Which figure is **not** a quadrilateral?

   A   B

   C   D

6. Which figure is a hexagon?

   A   B

   C   D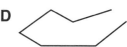

7. What polygon is used for each sign?

   _____   _____   _____

8. Answer *yes* or *no*. Explain.

   a. Is a circle a polygon?

   _____

   b. Can a polygon have more vertices than sides?

   _____

   _____

Objective 13.2: Define polygons and classify different types of polygons.

Lesson 13-3 Triangles

Model It

Words to Know In **equilateral triangles,** all sides are the same length and all angles have the same measure.

In **isosceles triangles,** exactly two sides have the same length and exactly two angles have the same measure.

Activity 1

Is the triangle an **isosceles triangle?**

If exactly two sides are equal, then the triangle is isosceles. Use a ruler.

$\overleftrightarrow{AB}$ and $\overleftrightarrow{AC}$ are each 2 cm long.

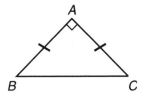

Practice 1

Circle the isosceles triangle.

Use a protractor. If exactly two angles are equal, then the triangle is isosceles. Mark the equal angles.

Activity 2

Is the triangle **equilateral?**

Use a ruler. If all sides are equal, then the triangle is equilateral.

Each side is 3 cm, so all sides are equal.

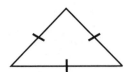

The triangle is equilateral.

Practice 2

Circle the equilateral triangle.

Use a protractor. If all angles are equal, then the triangle is equilateral.

On Your Own

Classify the triangle.

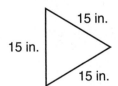

Write About It

Does it matter which way a triangle is turned? Explain why or why not.

Objective 13.3: Explore, compare, and classify different types of triangles.

Lesson 13-3 Triangles

Words to Know A **right triangle** has one right angle.

Example 1

The sum of a triangle's angle measures is 180°. Find the missing angle measure.

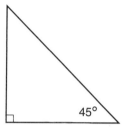

Add the angle measures: 90° + 45° = 135°.
Subtract from 180°: 180° − 135° = 45°.

The third angle measures 45°.

Practice 1

Find the missing angle measure. Show the operations.

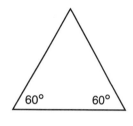

_____ + _____ = _____

_____ − _____ = _____

The third angle measures 60°

Example 2

Classify the triangle.

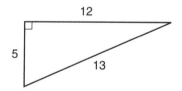

Examine the lengths of the sides.
No sides are equal.

Examine the measures of the angles.
There is 1 right angle.

The triangle is a **right triangle.**

Practice 2

Find the missing measures.
Then circle the right triangle.

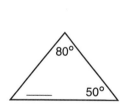 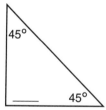

On Your Own

What do the other two measures on a right triangle have to add up to? Why?

Write About It

Can an equilateral triangle have a right angle?

Objective 13.3: Explore, compare, and classify different types of triangles.

Lesson 13-3 **Triangles**

1. Classify the triangle.

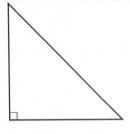

_____

2. Classify the triangle.

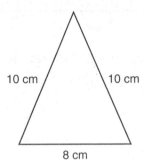

_____

3. Find the missing measure.

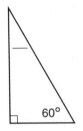

4. Which of the following best describes the triangle? Circe the letter of the correct answer.

  A equilateral   B right isosceles

  C right   D isosceles

5. Classify the triangle.

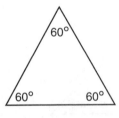

_____

6. Classify the triangle.

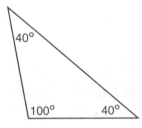

_____

7. Use a protractor. Label the angles. Then classify the triangle.

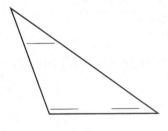

_____

8. Use a protractor. Label the angles. Then classify the triangle.

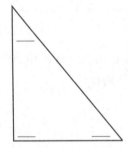

_____

Objective 13.3: Explore, compare, and classify different types of triangles.

Lesson 13-4 — Quadrilaterals

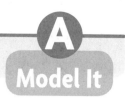
Model It

Words to Know   A **quadrilateral** is a 4-sided polygon.
Congruent means having equal measure.

Activity 1

Find a polygon tile that is a rectangle.

Find the length of each side.
Opposite sides have equal lengths.

Find the measure of each angle.
Each angle measures 90°.

A **rectangle** has congruent opposite sides and four 90° angles.

Practice 1

Find a square, a trapezoid, and a rhombus. Which of these is also a rectangle?

Look for _____ sides with equal lengths.

Look for angles that measure _____.

The _____ is also a rectangle.

Activity 2

Find a polygon tile that is a parallelogram.

Opposite sides are parallel.
Find the length of each side.
Opposite sides have equal lengths.

Find the measure of each angle.
Opposite angles have equal measures.
A **parallelogram** has two pairs of parallel sides. Also, opposite sides are congruent, and opposite angles are congruent.

Practice 2

Find a rectangle, rhombus, and trapezoid. Which shapes are also parallelograms?

Look for sides that are _____.

The _____ are parallelograms.

On Your Own

Fill in the blanks.

A square is both a _____

and a _____.

Write About It

Divide the rectangle into two triangles. Use them to explain why the sum of the angles in a quadrilateral is 360°.

Objective 13.4: Explore, compare, and classify different types of quadrilaterals.

Lesson 13-4 Quadrilaterals

B Understand It

Example 1

Classify the quadrilateral.

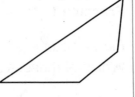

One pair is parallel. One pair is not. No sides are congruent.

A quadrilateral with one pair of parallel sides is a **trapezoid**. The shapes below are also trapezoids.

Practice 1

Classify the quadrilateral.

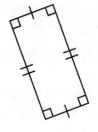

Which sides are parallel? _____

Which sides are congruent? _____

Which angles are congruent? _____

The quadrilateral is a _____

and a _____ .

Example 2

Find the measure of angle X.

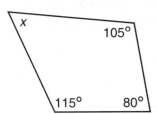

The sum of the angle measures is 360°. Write an equation.
$$105° + 80° + 115° + x = 360°$$
$$300° + x = 360°$$
$$300° - 300° + x = 360° - 300°$$
$$x = 60°$$
Angle X measures 60°.

Practice 2

Find the measure of angle X.

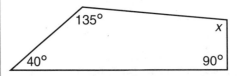

Write an equation.

____ + ____ + ____ + x = 360°

____ + x = 360°

____ − ____ + x = 360° − ____

x = ____ Angle X measures ____ .

On Your Own

Complete the following sentence in as many ways as possible. A square is also a ____ .

Write About It

Why can some quadrilaterals be classified in more than one way?

Objective 13.4: Explore, compare, and classify different types of quadrilaterals.

Lesson 13-4 Quadrilaterals

1. Classify the quadrilateral in as many ways as possible.

   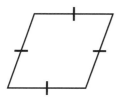

   _____

2. Classify the quadrilateral in as many ways as possible.

   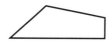

   _____

3. Which name **cannot** be used to describe the figure? Circle the letter of the correct answer.

   A quadrilateral B parallelogram

   C trapezoid D square

4. Which is **not** a quadrilateral? Circle the letter of the correct answer.

   A hexagon B trapezoid

   C square D rhombus

5. Sumita says this shape is a rectangle. Marvyn says that it is a square. Who is correct?

6. Find the measure of angle X.

   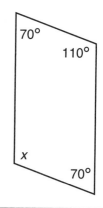

   _____

7. A figure has 4 congruent sides. The opposite sides are parallel. Can you classify the figure as a square?

8. Is the statement true? Explain.

   All quadrilaterals are rectangles.

Objective 13.4: Explore, compare, and classify different types of quadrilaterals.

Topic 13: Basic Geometric Figures

Topic Summary

Choose the correct answer. Explain how you decided.

1. Julius drew this set of lines on his paper. Identify the type of lines he drew.

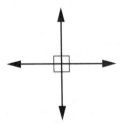

A acute

B parallel

C perpendicular

D supplemental

_____

_____

2. Which is the measure of angle *x*?

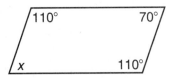

A 70°

B 90°

C 110°

D 35°

_____

_____

_____

Objective: Review basic geometric figures.

Topic 13 — Basic Geometric Figures

Mixed Review

1. Identify the value of the underlined digit.

   a. 7,0<u>6</u>5 _____

   b. <u>6</u>48 _____

   c. <u>9</u>,253 _____

   Volume 1, Lesson 2-4

2. Multiply.

   a. 521 × 6 _____

   b. 3 × 1,782 _____

   c. 4,236 × 5 _____

   Volume 4, Lesson 11-3

3. What is the measure of the angle?

   Volume 3, Lesson 10-4

4. A math teacher made 8 sets of 100 copies. How many copies did the math teacher make? Circle the letter of the correct answer.

   A 80 **B** 800

   C 8,000 **D** 80,000

   Volume 4, Lesson 11-1

5. Divide 56,000 ÷ 700. Circle the letter of the correct answer.

   A 800 **B** 80

   C 8,000 **D** 800,000

   Volume 4, Lesson 12-1

6. Explain why a square is a rectangle, but a rectangle is not a square.

   Volume 5, Lesson 13-4

Objective: Maintain concepts and skills.

Topic 14: Standard Measurement

Topic Introduction

Complete with teacher help if needed.

1. Find each answer.

   a. To change from feet to inches, multiply by _____.

   b. To change from feet to yards, divide by _____.

   c. To change from yards to inches, multiply by _____.

   Objective 14.1: Explore the basic units of length in the U.S. customary system.

2. Find each measure.

   a. 1 foot = _____ inches

   b. 1 yard = _____ feet

   c. 1 yard = _____ inches

   Objective 14.1: Explore the basic units of length in the U.S. customary system.

3. Find each measure.

   a. 1 cup = _____ fluid ounces

   b. 1 gallon = _____ quarts

   c. 1 quart = _____ pints

   d. 1 fluid ounce = _____ tablespoons

   Objective 14.3: Explore the basic U.S. customary units of liquid capacity.

4. Find each measure.

   a. 1 pound = _____ ounces

   b. 1 ton = _____ pounds

   Objective 14.2: Explore the basic U.S. customary units of weight.

Lesson 14-1: Basic Units of Length

Model It — A

Activity 1

Look at a ruler. How many **inches** are in one **foot**? 12 inches = 1 foot

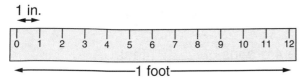

Would you measure the length of a chalkboard in inches or feet?

If the chalkboard is more than 12 inches long, **feet** are the best unit of measure.

Practice 1

Look at a yardstick. How many **feet** are in one **yard**? _____ = 1 yard

Would you use feet or yards to measure the length of a baseball field?

_____

Activity 2

Circle the unit that belongs in the blank.

My friend's dad is 6 _____ tall.

yards   inches   (feet)

Practice 2

Circle the unit that does **not** make sense in the blank.

My closet is 2 _____ wide.

yards   inches   feet

On Your Own

Circle the best measure for the length of a piece of paper.

feet   yards   inches

Write About It

Explain why it would be difficult to measure the length of the classroom in inches.

Objective 14.1: Explore the basic units of length in the U.S. customary system.

Level D

Lesson 14-1: Basic Units of Length

Understand It

Words to Know 1 foot = 12 inches   1 yard = 3 feet

Example 1

How many inches are in 3 feet?

There are 12 inches in 1 foot.
Make 3 groups of 12 inches.

←—12 in.—→ ←—12 in.—→ ←—12 in.—→

12 + 12 + 12 = 36
or   12 × 3 = 36
There are 36 inches in 3 feet.

Practice 1

How many feet are in 4 yards?

There are _____ feet in 1 yard.

Make 4 groups of _____ feet.

_____ + _____ + _____ + _____

or _____ × _____ = _____

There are _____ feet in 4 yards.

Example 2

How many feet are in 24 inches?

There are 12 inches in 1 foot.
Put the inches in groups of 12.

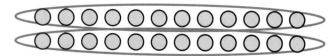

There are **2** feet in 24 inches.

Practice 2

How many yards are in 9 feet?

There are _____ feet in 1 yard.

Put the feet in groups of _____ .

There are _____ yards in 9 feet

On Your Own

Find the number of feet in 2 yards.

_____

Now compare.
2 yards ◯ 5 feet

Write About It

It is helpful to begin by asking whether the number in your answer will be larger or smaller than the number in the problem. Why?

Objective 14.1: Explore the basic units of length in the U.S. customary system.

Lesson 14-1 | **Basic Units of Length**

1. Would you use feet or inches to measure the length of your bed?

   _____

2. Would you use yards, feet, or inches to measure the length of a notebook?

   _____

3. Circle the most reasonable estimate for the height of a table.

   2 feet   2 inches   2 yards

4. Circle the most reasonable estimate for the length of a playground.

   8 feet   8 inches   8 yards

5. How many inches are in 4 feet?

   _____

6. How many yards are in 15 feet?

   _____

7. How many feet are in 36 inches?

   A 1   B 2

   C 3   D 4

8. Explain how to determine the number of inches in 5 feet and 2 inches.

Objective 14.1: Explore the basic units of length in the U.S. customary system.

Level D

Lesson 14-2 Basic Units of Weight

Model It

Activity 1

Put 5 nickels in one of the containers on the balance scale. Five nickels weigh about 1 ounce.

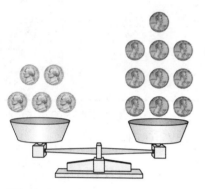

Find the number of pennies it takes to balance the scale.

10 pennies weigh about one ounce.

Practice 1

Put 5 nickels in one of the containers on the balance scale.

Find the number of dimes it takes to balance the scale.

_____ dimes weigh about one ounce.

Activity 2

Put 3 rolls of pennies in one of the containers on the balance scale. Three rolls of pennies weigh about 1 pound.

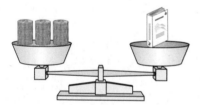

Find items in the room to balance the scale. A book balances the scale.

A book weighs about one pound.

Practice 2

Put 10 quarters in one of the containers on the balance scale. Ten quarters weigh about 2 ounces.

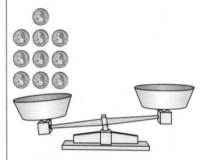

How many paper clips balance the scale?

_____ paper clips weigh 2 ounces.

On Your Own

Find something in the room that weighs about 4 ounces.

_____

Write About It

5 nickels weigh about one ounce. About how many nickels weigh 3 ounces? Explain.

Objective 14.2: Explore the basic U.S. customary units of weight

Lesson 14-2 — Basic Units of Weight

Understand It

Words to Know   16 **ounces** = 1 **pound**   2,000 **pounds** = 1 **ton**

Example 1

Choose the unit you would use to weigh each of the following. Use ounces, pounds, or tons.

The weight of an elephant is measured in <u>tons</u>.

The weight of a handful of paper clips is measured in <u>ounces</u>.

Practice 1

Choose the unit you would use to weigh each of the following. Use ounces, pounds, or tons.

The weight of a dog is measured in _____.

The weight of a piece of bread is measured in _____.

Example 2

Circle the items that weigh more than one pound.

14 ounces   12 ounces   17 ounces

1 pound = 16 ounces

17 ounces is greater than 16 ounces.

Practice 2

Circle the items that weigh less than one pound.

18 ounces   14 ounces   11 ounces

On Your Own

What unit would you use to weigh a caterpillar?

_____

Write About It

Why wouldn't ounces be the best unit of measure for the weight of a car?

Objective 14.2: Explore the basic U.S. customary units of weight.

Lesson 14-2   **Basic Units of Weight**

Try It

1. 10 pennies weigh about one ounce. 40 pennies weigh about how many ounces?

   _____

2. Carrie placed 3 pounds of carrots on a serving platter. The platter weighs one pound. How much do the platter and carrots weigh in all?

   _____

3. Which unit of measure would you use to weigh a stack of textbooks?

   A ounces   B pounds

   C inches   D tons

4. Which is the best estimate for the weight of an apple?

   A 4 ounces   B 15 ounces

   C 1 pound   D 2 pounds

5. Circle the items that weigh more than one pound.

   12 ounces   19 ounces   13 ounces

6. List the measurements in order from lightest to heaviest.

   1 ton, 20 ounces, 2 pounds, 5 ounces

   _____

   _____

7. A horse weighs 1,200 pounds. Does the horse weigh more or less than one ton?

   _____

8. Two different objects are the same size. Does this mean that they weigh the same amount? Explain.

Objective 14.2: Explore the basic U.S. customary units of weight.

Lesson 14-3: Basic Units of Liquid Capacity

Model It

Words to Know A **fluid ounce** is used to measure the amount of liquid, but an **ounce** is used to measure weight.

Activity 1

Fill a measuring cup to the 1-cup line.

Turn the measuring cup around to see how many fluid ounces are in one cup.

1 cup = 8 fluid ounces

Practice 1

Fill a pint container with water.

Now pour the water into the liquid measuring cup.

How many cups are in one pint?

1 pint = _____ cups

Activity 2

You will need a pint and quart container.

Fill your pint container with water and pour it into the quart container.

How many pints does it take to fill a quart?

1 quart = 2 pints

Practice 2

You will need a quart and gallon container.

Estimate: How many quarts do you think it takes to fill a gallon? _____

Fill your quart container and empty it into the gallon. Repeat until the gallon container is full.

How many quarts did it take to make a gallon? _____

1 gallon = _____ quarts

On Your Own

Which is greater: 3 cups or 1 pint?

_____

Write About It

How many pints are in a gallon? Explain how you determined your answer.

Objective 14.3: Explore the basic U.S. customary units of liquid capacity.

Level D

Lesson 14-3 | **Basic Units of Liquid Capacity**

Words to Know
8 fluid ounces = 1 cup
2 cups = 1 pint
2 pints = 1 quart
4 quarts = 1 gallon

Example 1

Choose the unit that you would use to measure each of the following. Choose from cups, pints, quarts, or gallons.

Soup in a bowl is measured in <u>cups</u>.

Water in a bathtub is measured in <u>gallons</u>.

Practice 1

Choose the unit that you would use to measure each of the following. Choose from fluid ounces, cups, pints, quarts, or gallons.

Water in a washing machine is measured in _____.

Yogurt in a small container is measured in _____.

Example 2

Find the number of quarts in 2 gallons.

There are 4 quarts in 1 gallon, so multiply 4 quarts by 2.

4 × 2 = 8

There are 8 quarts in 2 gallons.

Practice 2

Find the number of pints in 10 cups.

Each pint holds _____ cups.

How many groups of _____ are in 10?

10 ÷ _____ = _____

There are _____ pints in 10 cups.

On Your Own

Show how to find the number of gallons in 16 quarts.

_____
_____
_____
_____

Write About It

Henry says there are more quarts than cups in 2 gallons. Do you agree? Explain.

Objective 14.3: Explore the basic U.S. customary units of liquid capacity.

Lesson 14-3 Basic Units of Liquid Capacity

1. Find the number of fluid ounces in 2 pints.

   _____

2. 6 pints is the same as how many quarts?

   _____

3. Solve.

   a. _____ cups = 24 fluid ounces

   b. 4 quarts = _____ pints

   c. 1 gallon = _____ cups

4. Which unit of measure would you use to determine the amount of water in a fish pond? Circle the letter of the best answer.

   A fluid ounces   B pints

   C quarts   D gallons

5. List in order from least to greatest:

   1 pint, 1 cup, 3 cups, 10 fluid ounces

   _____

   _____

   _____

6. Which unit of measure would be used to determine the amount of juice in a juice box?

   A fluid ounces   B gallons

   C quarts   D tons

7. Darla poured 2 gallons of punch into a punch bowl. Then she added 3 quarts of punch. How many total quarts of punch are in the bowl?

   _____

8. Explain why cups are **not** the best unit of measure to determine the water in a swimming pool.

   _____

   _____

   _____

   _____

Objective 14.3: Explore the basic U.S. customary units of liquid capacity.

Level D

Topic 14 — Standard Measurement

Topic Summary

Choose the correct answer. Explain how you decided.

1. There are 432 inches from Malcolm's front door to his mailbox. How many yards does Malcolm have to walk from his front door to his mailbox?

   A 12 yards

   B 36 yards

   C 15,552 yards

   D 5,184 yards

   _____
   _____
   _____

2. The bakery has two bags of flour. One bag weighs 32 ounces. The other bag weighs 6 pounds. How many pounds of flour does the bakery have?

   A 38 ounces

   B 8 pounds

   C 38 pounds

   D 14 pounds

   _____
   _____
   _____

Objective: Review measurement conversions.

Topic 14

Standard Measurement

Mixed Review

1. Find each sum or difference mentally.

   a. 8 + 7 = _____   b. 3 + 5 = _____

   c. 20 − 6 = _____   d. 15 − 6 = _____

   e. 12 − 7 = _____   f. 10 + 4 = _____

   g. 9 + 8 = _____   h. 11 − 5 = _____

   i. 17 − 10 = _____   j. 6 + 8 = _____

   Volume 2, Lesson 5-1

2. Write each number in words.

   a. 408

   _____

   b. 126

   _____

   Volume 1, Lesson 2-2

3. Find each product mentally.

   a. 6 × 4 = _____   b. 9 × 7 = _____

   c. 3 × 5 = _____   d. 8 × 4 = _____

   e. 2 × 7 = _____   f. 8 × 6 = _____

   g. 9 × 3 = _____   h. 4 × 7 = _____

   Volume 4, Lesson 9-5

4. For his trail mix, Devon bought 16 ounces of dried pineapples, 12 ounces of banana chips, 12 ounces of dried papaya, and 8 ounces of cashews. How many pounds of trail mix did he make?

   Volume 5, Lesson 14-2

5. Divide.

   a. 7 ÷ 7 = _____

   b. 7 ÷ 1 = _____

   c. 0 ÷ 7 = _____

   Volume 4, Lesson 10-3

6. Use doubles to solve.

   a. 5 + 6 = _____

   doubles fact: _____

   b. 9 + 8 = _____

   doubles fact: _____

   Volume 2, Lesson 3-3

Level D

Topic 15: Metric Measurement

Topic Introduction

Complete with teacher help if needed.

1. Using the metric system of measurement, write the most reasonable unit of measure for each.

   a. the weight of a dog _____

   b. the capacity of a juice box _____

   c. the weight of a pencil _____

   d. the capacity of water needed to fill a kitchen sink _____

2. Find each measure.

   a. 1 meter = _____ centimeters

   b. 1 meter = _____ millimeters

   c. 1 kilometer = _____ meters

Objective 15.3: Explore the basic metric units of mass.
Objective 15.4: Explore the basic metric units of capacity.

Objective 15.2: Explore the basic metric units of length.

3. Find each answer.

   a. To change from meters to centimeters, multiply by _____.

   b. To change from millimeters to meters, divide by _____.

   c. To change from kilometers to meters, multiply by _____.

4. Which is greater?

   a. 10 mL or 1 liter _____

   b. 1,000 g or 2 kg _____

   c. 12 mm or 1 cm _____

   d. 200 cm or 1 m _____

Objective 15.2: Explore the basic metric units of length.

Objective 15.1: Explore the b...

Lesson 15-1 — Basic Metric Prefixes

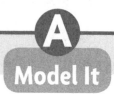
Model It

Words to Know Milli- means one-thousandth. Centi- means one-hundredth.
Kilo- means one thousand.

1 meter = 1,000 millimeters	1 **meter** = 100 centimeters	1,000 meters = 1 kilometer
1 gram = 1,000 milligrams	1 **gram** = 100 centigrams	1,000 grams = 1 kilogram
1 liter = 1,000 milliliters	1 **liter** = 100 centiliters	1,000 liters = 1 kiloliter

Activity 1

Circle the smaller amount.

a. (1 millimeter)   1 kilometer

b. (1 centigram)   1 gram

Practice 1

Circle the smaller amount.

a. 1 meter   1 centimeter

b. 1 gram   1 milligram

Activity 2

Circle the smaller amount.

a. (105 centimeters)   1 kilometer

b. (100 grams)   1 kilogram

Practice 2

Circle the smaller amount.

a. 3 milligrams   30 grams

b. 2 liters   20 centiliters

On Your Own

Order from least to greatest.

50 meters, 50 centimeters, 5 kilometers

_____

_____

Write About It

Suppose you were converting grams to kilograms. Would the number of kilograms be greater or less than the number of grams? Explain.

Objective 15.1: Explore the basic metric prefixes.

Lesson 15-1 — Basic Metric Prefixes

Understand It

1 meter = 1,000 millimeters	1 **meter** = 100 centimeters	1,000 meters = 1 kilometer
1 gram = 1,000 milligrams	1 **gram** = 100 centigrams	1,000 grams = 1 kilogram
1 liter = 1,000 milliliters	1 **liter** = 100 centiliters	1,000 liters = 1 kiloliter

Example 1

Fill in the table to show equivalent amounts.

liters	1	2	3	4
milliliters	1,000	2,000	3,000	4,000

Practice 1

Fill in the table to show equivalent amounts.

meters	1	2	___
centimeters	100	___	300

Example 2

Fill in the table to show equivalent amounts.

centigrams	100	500	800	1,000
grams	1	5	8	10

Practice 2

Fill in the table to show equivalent amounts.

liters	1,000	___	6,000
kiloliters	1	4	___

On Your Own

Fill in the blanks.

5,000 millimeters = _____ meters

2 kiloliters = _____ liters

Write About It

What is the difference between a millimeter and a kilometer?

Objective 15.1: Explore the basic metric prefixes.

Lesson 15-1 — Basic Metric Prefixes

Try It

1. Fill in the blanks to show equivalent amounts.

grams	___	2,000	3,000
kilograms	1	___	3

2. 100 centimeters is equivalent to which of the following? Circle the correct answers.

A 1 millimeter   B 1 kilometer

C 1,000 millimeters   D 1 meter

3. Compare. Use <, >, or =.

a. 3,000 milliliters ___ 3 kiloliters

b. 200 centigrams ___ 2 grams

4. Circle the smaller unit in each pair.

a. meter   kilometer

b. milligram   centigram

5. Fill in the table to show equivalent amounts.

grams	1	5	10
milligrams	1,000	___	___

6. Fill in the table to show equivalent amounts.

liters	1	6	___
centiliters	100	___	1,200

7. Are any of these amounts equivalent? Explain.

1,000 centimeters, 10 meters, 1 kilometer

8. A tree measures 8 meters tall. Can you write this height as 8,000 centimeters? Why or why not?

Objective 15.1: Explore the basic metric prefixes.

Lesson 15-2: Basic Metric Units of Length

Activity 1

Measure the line segment to the nearest centimeter.

Align the end of the centimeter ruler with the end of the line segment.

The length of the line segment to the nearest centimeter is 7 cm.

Practice 1

Measure the line segment to the nearest centimeter.

•————————•

The length of the line segment to the nearest centimeter is _____ cm.

Activity 2

Measure the length of the paper clip to the nearest millimeter.

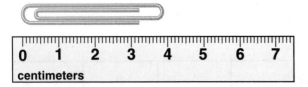

Line up the end of the centimeter ruler with the end of the line segment.

The length of the line segment to the nearest millimeter 39 mm.

Practice 2

Measure the length of the shoe to the nearest millimeter.

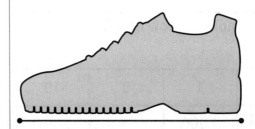

The length of the line segment to the nearest millimeter _____ mm.

On Your Own

Measure the length of the box to the nearest millimeter.

Write About It

Explain how to measure to the nearest centimeter.

Objective 15.2: Explore the basic metric units of length.

Lesson 15-2: Basic Metric Units of Length

Understand It — B

Example 1

Choose the metric unit that you would use to measure each of the following. Use mm, cm, m, or km.

The length of a calculator would be measured in <u>centimeters (cm)</u>.

The length of a soccer field would be measured in <u>meters (m)</u>.

Practice 1

Choose the metric unit that you would use to measure each of the following. Use mm, cm, m, or km.

The distance between the Earth and the moon would be measured in _____.

The length of a ladybug would be measured in _____.

Example 2

a. How many cm are in 60 mm?

There are 10 mm in 1 cm, so divide 60 by 10. 60 ÷ 10 = 6. There are 6 cm in 60 mm.

b. How many mm are in 8 cm?

There are 10 mm in 1 cm, so multiply 8 by 10. 8 × 10 = 80. There are 80 mm in 8 cm.

Practice 2

a. How many cm are in 100 mm?

There are 10 mm in 1 cm, so divide 100 by _____. 100 ÷ _____ = _____. There are _____ cm in 100 mm.

b. How many mm are in 7 cm?

There are 10 mm in 1 cm, so multiply _____ by 10. 7 × _____ = _____. There are _____ mm in 7 cm.

On Your Own

Measure the length of your thumb to the nearest cm.

_____

Write About It

20 is larger than 3. However, a 20-mm piece of string is shorter than a 3-cm piece of string. Explain why.

Objective 15.2: Explore the basic metric units of length.

Lesson 15-2 — Basic Metric Units of Length

Try It

1. Measure the line segment to the nearest centimeter.

   •————————————————•

   _____

2. Draw a line segment that has a length of 57 mm.

3. What is the length of the camera to the nearest centimeter?

   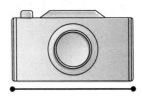

   A 2 cm   B 3 cm

   C 20 cm   D 30 cm

4. Which metric unit would you use to measure the length of a swimming pool?

   A gallon   B centimeter

   C meter   D kilometer

5. Which of the following rounds to 30 cm when rounded to the nearest cm?

   A 32 cm   B 33 mm

   C 37 cm   D 28 mm

6. Suppose these two segments are connected to form a straight line. What would be the total length in mm?

   •————4 cm————•   •—8 mm—•

   _____

7. How many cm are in 50 mm?

   _____

8. Explain why a meter is **not** the best unit to measure the length of an envelope.

   _____

   _____

Objective 15.2: Explore the basic metric units of length.

Lesson 15-3 | **Basic Metric Units of Mass**

Words to Know Mass is the amount of matter in an object.

Activity 1

Put the 10-gram bar in one of the containers on the Rocker Balance Scale.

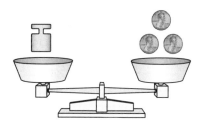

Find the number of pennies it takes to balance the scale. About three pennies balance the scale.

3 pennies have a mass of about 10 grams.

Practice 1

Put the 5-gram bar in one of the containers on the balance scale.

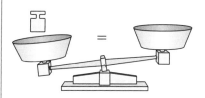

Find the number of dimes it takes to balance the scale.

_____ dimes have a mass of about 5 grams.

Activity 2

Put the 20-gram bar in one of the containers on the Rocker Balance Scale.

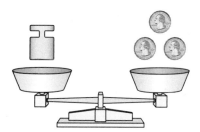

Find the number of quarters it takes to balance the scale.

3 quarters have a mass of about 20 grams.

Practice 2

Put the 50-gram bar in one of the containers on the balance scale.

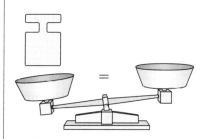

Find the number of nickels it takes to balance the scale.

_____ nickels have a mass of about 50 grams.

On Your Own

Find something in the room that has a mass of about 5 grams.

Write About It

Explain how to use the scale to find an object that has a mass of 4 grams.

Objective 15.3: Explore the basic metric units of mass.

Lesson 15-3 — Basic Metric Units of Mass

Understand It

Words to Know   1,000 **grams** = 1 **kilogram**

Example 1

Circle the mass of the wooden baseball bat.

2 grams or (2 kilograms)

The mass of a baseball bat is about 2 kilograms.

Practice 1

Circle the mass of the toothbrush.

8 grams or 8 kilograms

The mass of a toothbrush is about _____ grams.

Example 2

Choose the unit you would use to measure each of the following. Use grams or kilograms.

The mass of a cat is measured in <u>kilograms</u>.

The mass of a feather is measured in <u>grams</u>.

Practice 2

Choose the unit you would use to measure each of the following. Use grams or kilograms.

The mass of a goldfish is measured in _____.

The mass of a pencil is measured in _____.

On Your Own

Circle the mass of a bag of flour.

2 grams or 2 kilograms

Write About It

Explain how to find the number of grams in 5 kilograms.

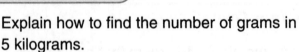

Objective 15.3: Explore the basic metric units of mass.

Lesson 15-3   **Basic Metric Units of Mass**   Try It

1. 3 pennies have a mass of about 10 grams. What is the mass of 6 pennies?

   _____

2. 10 nickels have a mass of about 50 grams. How many nickels have a mass of 25 grams?

   _____

3. Which unit should you use to measure the mass of a watermelon?

   A ounce   B gram

   C meter   D kilogram

4. Which is the best estimate for the mass of an orange?

   A 14 grams   B 14 kilograms

   C 141 grams   D 141 kilograms

5. Circle the mass of the car.

   600 grams or 600 kilograms

6. Circle the mass of the mouse.

   15 grams or 15 kilograms

7. 7,000 grams = _____ kilograms

   A 7   B 70

   C 7,000   D 7,000,000

8. Explain how to use the scale to find an object that has a mass of 35 grams.

   _____

   _____

   _____

   _____

   _____

Objective 15.3: Explore the basic metric units of mass.

Level D

Lesson 15-4: Basic Metric Units of Capacity

Words to Know One **liter (L)** is the same as 1,000 **milliliters (mL)**.

Activity 1

a. How many milliliters are in 3 liters?
There are <u>1,000</u> milliliters in 1 liter.
Multiply 3 by 1,000.
3 × 1,000 = 3,000

There are 3,000 milliliters in 3 liters.

b. How many liters are in 5,000 milliliters?
There is <u>1</u> liter in 1,000 milliliters.
Divide 5,000 by 1,000.
5,000 ÷ 1,000 = 5.

There are 5 liters in 5,000 milliliters.

Practice 1

a. How many milliliters are in 6 liters?
There are _____ in 1 liter.
Multiply 6 by _____.
6 × _____ = _____

There are _____ milliliters in 6 liters.

b. How many liters are in 4,000 milliliters?
There is _____ liter in 1,000 milliliters.
Divide 4,000 by _____.
4,000 ÷ _____ = _____

There are _____ liters in 4,000 milliliters.

Activity 2

Which is greater: 3 liters or 3,500 milliliters?

Convert liters to milliliters.

3 liters × 1,000 = 3,000 milliliters

Compare 3,000 milliliters and 3,500 milliliters.

3,500 milliliters is greater than 3 liters.

Practice 2

Which is less: 4 liters or 3,000 milliliters?

Convert milliliters to liters.

3,000 milliliters ÷ _____ = _____ liters

Compare 4 liters and 3 liters.

_____ is less than _____.

On Your Own

How many milliliters are in a 2-liter bottle?

_____

Write About It

How can you convert liters to milliliters?

Objective 15.4: Explore the basic metric units of capacity.

Lesson 15-4: Basic Metric Units of Capacity

B Understand It

Example 1

Circle the capacity of the eye dropper.

(1 milliliter) or 1 liter

The capacity of the eye dropper is 1 milliliter.

Practice 1

Circle the capacity of the mug.

300 milliliters or 300 liters

Example 2

Choose the unit you would use to measure each of the following. Choose from milliliters or liters.

The lemonade in a jug can be measured in <u>liters</u>.

A drop of water is measured in <u>milliliters</u>.

Practice 2

Choose the unit you would use to measure each of the following. Choose from milliliters or liters.

The gasoline in a gas tank can be measured in _____.

A dose of liquid medicine is measured in _____.

On Your Own

Circle the capacity of the watering can.

4 milliliters or 4 liters

Write About It

Marcia was so thirsty that she drank 30 milliliters of water! Explain why this sentence does not make sense.

Objective 15.4: Explore the basic metric units of capacity.

Lesson 15-4: Basic Metric Units of Capacity

1. Find the number of liters in 10,000 milliliters.
   _____

2. Find the number of milliliters in 8 liters.
   _____

3. Circle the capacity of the spoon.

   4 milliliters or 4 liters

4. Circle the capacity of the bucket.

   6 milliliters or 6 liters

5. 5,000 milliliters = _____ liters

   A 5
   B 50
   C 5,000
   D 5,000,000

6. Determine the sum of 4 liters, 200 milliliters, and 1 liter.
   _____

7. Carlos poured a 1-liter bottle of tea equally into 4 separate cups. How much tea is in each cup?
   _____

8. Laura wants to fill a fishbowl with 3 liters of water. She has a container which holds 500 milliliters. Explain how she can use the container to fill the fishbowl.
   _____
   _____

Objective 15.4: Explore the basic metric units of capacity.

Topic 15: Metric Measurement

Topic Summary

Choose the correct answer. Explain how you decided.

1. Sylvia wants to measure the length of her foot. Which metric unit would be the most reasonable measure to use?

   A millimeter

   B milliliter

   C centimeter

   D meter

   _____
   _____
   _____
   _____
   _____

2. Kentrall is making a fruit juice punch for his party. If he makes two liters of punch, how many milliliters of juice does he need?

   A 2

   B 20

   C 2,000

   D 200

   _____
   _____
   _____

Objective: Review the metric system of measurement.

Level D

Topic 15: Metric Measurement

Mixed Review

1. Find each product mentally.

a. 9 × 6 = _____   b. 6 × 7 = _____

c. 3 × 8 = _____   d. 10 × 4 = _____

e. 7 × 4 = _____   f. 5 × 9 = _____

g. 8 × 4 = _____   h. 8 × 7 = _____

i. 9 × 10 = _____   j. 6 × 8 = _____

Volume 4, Lesson 9-5

2. Give the digit in the indicated place.

3,690

a. tens _____

b. thousands _____

c. hundreds _____

d. ones _____

Volume 1, Lesson 2-4

3. Find each sum.

a. 468 + 831 = _____

b. 236 + 765 = _____

c. 87 + 538 = _____

d. 770 + 821 = _____

Volume 2, Lesson 6-1

4. Which of the following is NOT a polygon? Circle the letter of the correct answer.

A triangle   B circle

C parallelogram   D octagon

Volume 5, Lesson 13-2

5. Make a drawing of base ten blocks to represent the number 600 + 30 + 9. Then write the standard form for the number.

Volume 1, Lesson 2-2

6. Draw a picture to represent 4 × 2.

Volume 4, Lesson 9-3

Objective: Maintain concepts and skills.

Topic 16: Measures Geometric Figures

Topic Introduction

Complete with teacher help if needed.

1.

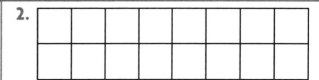

a. Between which two inch marks is the length of the paper clip?

_____

b. Is the length of the paper clip more or less than halfway between the two measures?

_____

c. Is the length closer to 1 in. or 2 in.?

_____

Objective 16.1: Measure the length of an object to the nearest inch or centimeter.

2.

a. Each small square stands for 1 square inch. How many square inches are in each row? _____

b. How many rows are there? _____

c. How many square inches are there all together? _____

Objective 16.3: Estimate or determine the area of figures by covering them with squares.

3.

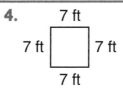

3 cm
1 cm
4 cm

a. How many centimeter cubes are in the bottom layer? _____

b. How many layers will fit inside? _____

c. How many centimeter cubes will fill the box? _____

Objective 16.5: Estimate or determine the volume of solid figures by counting the number of cubes that would fill them.

4.

7 ft
7 ft   7 ft
7 ft

a. What are the lengths of the 4 sides of the figure?

_____

b. What is the total distance around the figure?

_____

Objective 16.2: Find the perimeter of a polygon with integer sides.

Lesson 16-1: Length

Words to Know An **inch** is a unit of length in the U.S. system.
A **centimeter** is a unit of length in the metric system.

Activity 1

Determine which inch mark on the ruler is closest to the length. The paper clip is longer than 1 inch, but it is shorter than 2 inches. To the nearest inch, the length of the paper clip is 2 inches.

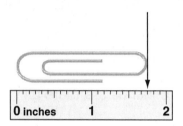

Practice 1

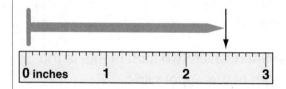

The nail's length is between _____ inches and _____ inches.

Since it meets the halfway mark, it is closer to _____ inches.

Activity 2

The leaf-shaped pin measures more than 3 centimeters, but less than 4 centimeters. It is closer to 3 centimeters. To the nearest centimeter, the pin is 3 centimeters long.

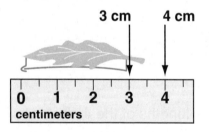

Practice 2

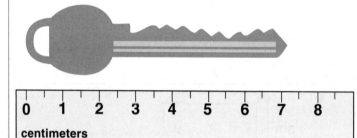

The key is closer to which centimeter mark? _____

On Your Own

Give the length of the line segment to the nearest inch and to the nearest centimeter.

_____

_____

Write About It

A line segment is between 3 inches and 4 inches long. Which measurement describes the length to the nearest inch? Explain.

Objective 16.1: Measure the length of an object to the nearest inch or centimeter.

Lesson 16-1: Length

Understand It — B

Example 1

Draw a line that is 2 inches long to the nearest inch. Use a ruler.

_____

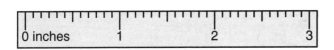

Practice 1

Draw a line that is 5 centimeters long to the nearest centimeter. Use a ruler.

Example 2

Draw a line 1 inch long. About how many centimeters is it?

_____

The line is about 3 centimeters long.

Practice 2

Draw a line 10 centimeters long. About how many inches is it?

The line is about _____ inches long.

On Your Own

Measure this eraser. About how many centimeters is it?

Write About It

Joy measures this lip balm and says it measures 2 inches to the nearest inch. Is she right? Why or why not?

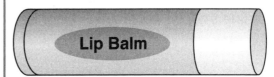

Objective 16.1: Measure the length of an object to the nearest inch or centimeter.

Lesson 16-1 — Length

Try It

1. Give the length of each line segment to the nearest inch.

   a.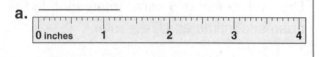

   b.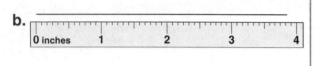

2. Give the length of each line segment to the nearest centimeter.

   a.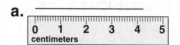

   b.

3. Draw a segment with a length you would round to the following measures.

   a. down to 1 inch

   b. up to 4 centimeters

4. Draw a line that is 2 inches long. Use a ruler.

5. Draw a line that is 7 centimeters long. Use a ruler.

6. About how long is this toy car?

   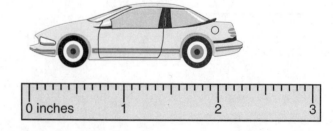

Objective 16.1: Measure the length of an object to the nearest inch or centimeter.

Lesson 16-2: Perimeter

Model It

Words to Know A **polygon** is a two-dimensional geometric shape that is closed and has segments for all of its sides.
The **perimeter** of a polygon is the distance around all its sides.

Activity 1

The perimeter is found by counting all the 1-cm squares as you move around the figure. Start at any corner and follow the arrows.

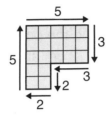

perimeter = (5 + 3 + 3 + 2 + 2 + 5) cm
= 20 cm

Practice 1

Write the length of each side in inches.

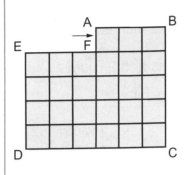

perimeter = _____

Activity 2

Find the perimeter of a square in which each side measures 3 meters.

number of sides in a square = 4

each side's measurement = 3 m

perimeter = 3 m + 3 m + 3 m + 3 m
= 4 sides × 3 m
= 12 m

Practice 2

Find the perimeter of a triangle in which each side measures 5 in.

number of sides in a triangle = _____

each side's measurement = _____

perimeter = _____

On Your Own

A four-sided polygon has a perimeter of 50 cm. Three sides have the same length. Give possible measurements for each of the sides.

Write About It

The perimeter of a rectangle is 16 inches, and each long side measures 5 inches. Explain how to find the length of each short side.

16-2: Find the perimeter of a polygon with integer sides.

Lesson 16-2 Perimeter

B Understand It

Words to Know The sides of a **regular polygon** are all the same length.
A **pentagon** is a polygon with five sides.
A **hexagon** is a polygon with six sides.
An **octagon** is a polygon with eight sides.

Example 1

Find the perimeter of a rectangle with a length of 6 inches and width of 2 inches.

A rectangle has 2 lengths and 2 widths.

length = 6 in.
width = 2 in.   width = 2 in.
length = 6 in.

perimeter = [(2 × length) + (2 × width)] inches
= [(2 × 6) + (2 × 2)] inches
= [12 + 4] inches
= 16 in.

Practice 1

Find the perimeter of a rectangle with a length of 12 cm and a width of 7 cm.

Label the sides of the rectangle.

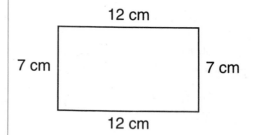

perimeter = [(2 × _____) + (2 × _____)] cm

= [_____ + _____] cm

= _____ cm

Example 2

Find the perimeter of a mat shaped like a regular hexagon whose sides measure 10 in.

perimeter = 6 sides × 10 in. per side = 60 in.

Practice 2

Find the perimeter of a regular octagon whose sides measure 11 m.

_____

On Your Own

A regular (equilateral) triangle and a regular pentagon each have a perimeter of 15 inches. How long is each side of each polygon?

Write About It

The perimeter of a hexagon is 24 ft. If each side is increased by 4 ft, what is its new perimeter? Explain how to find the answer.

Objective 16.2: Find the perimeter of a polygon with integer sides.

Lesson 16-2   **Perimeter**

Try It

1. Find the perimeter of each figure.

   a. _____

   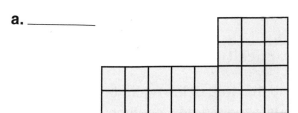

   b. _____

   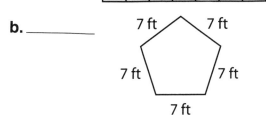

   c. _____

   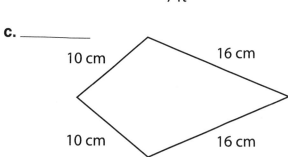

2. Determine the perimeter of each polygon.

   a. a triangle with sides measuring 12 ft, 8 ft, and 19 ft

   _____

   b. a rectangle with a length of 9 m and a width of 6m

   _____

   c. regular hexagon with each side measuring 8 cm

   _____

3. Answer each of the following for a rectangle with a length of 10 m and a width of 4 m.

   a. What is its perimeter? _____

   b. What would its perimeter be if both the length and the width were doubled?

   _____

4. What is the perimeter of a square with 7-inch sides? Circle the letter of the correct answer.

   A 28 in.   B 77 in.

   C 49 in.   D 14 in.

5. A gazebo is in the shape of a regular octagon. Each side measures 10 feet. Clarice puts lattice around the top of the gazebo. How many feet of lattice does she use?

   _____

6. A rectangular room is 14 feet long and 12 feet wide. Que placed a trim at the top of the walls around the room. How many feet of trim did she use?

   _____

Objective 16.2: Find the perimeter of a polygon with integer sides.

Lesson 16-3 Area

Model It

Words to Know Area is the amount of surface inside a 2-dimensional shape. **Area** is measured in **square units**, such as square inches (sq in.) or square centimeters (sq cm).

Activity 1

Find the area of the figure below if each small square measures 1 inch on each side.

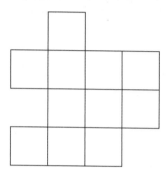

Count the square inches to find the area. The area is 11 square inches.

Practice 1

Find the area of the figure below if each small square measures 1 centimeter on each side.

The area is _____ _____ _____.

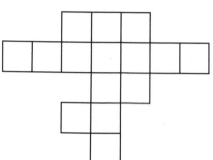

Activity 2

The rectangle is 3 units long and 2 units wide. Its area is 6 square units.

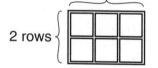

3 square units per row × 2 rows = 6 square units

Practice 2

The rectangle is 5 units long and 3 units wide. Draw lines to show how to divide the rectangle into 1-unit squares.

What is the area of the rectangle?

_____

On Your Own

The area of a rectangle is 18 square units. What are a possible length and width for the rectangle?

Write About It

Jan has a rectangle that is 6 inches long and 4 inches wide. How can she find the area of the rectangle into square inches?

Objective 16.3: Estimate or determine the area of figures by covering them with squares.

Lesson 16-3 Area — Understand It (B)

Example 1

What is the total area of all six faces?

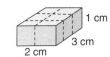

front + back = ▭ + ▭ = 4 sq cm
left + right = ▭ + ▭ = 6 sq cm
top + bottom = ▭ + ▭ = 12 sq cm

total area: (4 + 6 + 12) sq cm = 22 sq cm

Practice 1

Use grid paper to draw each face. Count the squares to find the total area of all faces.

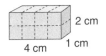

front + back = _____

left + right = _____

top + bottom = _____

total area: _____

Example 2

The figure has been covered with 1-inch squares. One estimate of its area is 16 in² because there are 15 whole squares and 4 partial squares in the figure.

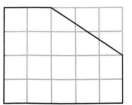

Practice 2

The figure to the right has been covered with 1-centimeter squares.

Give an estimate of the area of the figure.

_____

On Your Own

Use the centimeter grid. Draw a figure that is **not** a square or rectangle, and estimate its area.

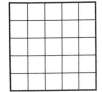

Write About It

The front of a rectangular box is 8 sq cm, the left face is 6 sq cm, and the top is 12 sq cm. Explain how to find the total area of the box.

Objective 16.3: Estimate or determine the area of figures by covering them with squares.

Lesson 16-3 Area

Try It

1. The grid paper shows square inches. Estimate the area of the figure.

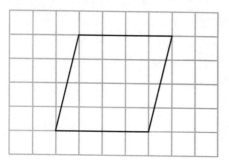

_____

2. The grid paper shows square meters. What is the area of the figure?

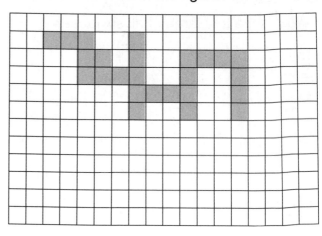

3. The grid paper shows square centimeters. What is the area of this rectangle?

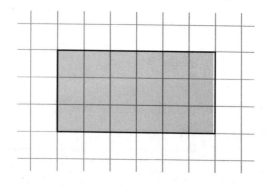

_____

4. What is the total area of this prism?

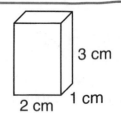

A 44 sq cm

B 22 sq cm

C 11 sq cm

D 6 sq cm

5. Each small square measures one square foot. What is the total area of each of these boxes in square feet?

a.   b.

_____   _____

6. Juan needs to find the area of a wall that has a width of 15 feet and a length of 10 feet. What is the area of the wall?

_____

7. Draw a figure that has an area of 20 square units.

Objective 16.3: Estimate or determine the area of figures by covering them with squares.

Lesson 16-4 Area of Rectangles

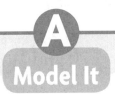
Model It

Words to Know The **area** of a figure is the number of square units needed to cover its surface. A **square unit** is a unit of area with dimensions 1 unit by 1 unit.

Activity 1

To find the area of the rectangle, use:
Area = Length × Width.
Area = 3 × 5 = 15

The area of the rectangle is 15 cm².

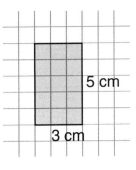

Practice 1

Draw a rectangle on centimeter grid paper. Find the area. What is the area of your rectangle?

_____

Activity 2

Mr. Collins measured his bulletin board. He uses a yardstick to measure the length and width.

Area = Length × Width

Area = 8 × 5 = 40

The area of the bulletin board is 40 ft².

Practice 2

Find the area of a bulletin board and the area of a sheet of notebook paper. Choose two different units of measure. Measure each object to the nearest whole number.

_____
_____
_____
_____

On Your Own

Find the area of a rectangle with length of 4 meters and a width of 16 meters.

Write About It

Why did you choose the units of measure you used to find the area of the bulletin board and the area of the sheet of paper in Practice 2?

Objective 16.4: Measure the area of rectangular shapes by using appropriate units.

Lesson 16-4 Area of Rectangles

B Understand It

Example 1

What would be appropriate units of measure to describe the area of Texas?

The state has a very large area. You would want to use a large unit.

Either square miles or square kilometers would be an appropriate unit of measure.

Practice 1

A company had a sign painted on one half of a rectangular-shaped roof of a barn. What would be appropriate units of measure for the area of the sign? Explain the reason for your choice.

_____
_____
_____
_____
_____
_____

Example 2

What is the area of the welcome mat?   21 in.   *Welcome*   32 in.

Area = length × width

Area = 32 × 21 = 672 in²

The area of the welcome mat is 672 in².

Practice 2

Serena hung a mirror on the wall. What is the area of the wall that is covered by the mirror?   60 cm   mirror   40 cm

_____

On Your Own

Eli bought a rug in the shape of a rectangle. It is 8 feet wide and 10 feet long. What is the area of the rug?

_____

Write About It

Would it be more appropriate to measure the area of an envelope in square centimeters or square meters? Explain.

Objective 16.4: Measure the area of rectangular shapes by using appropriate units.

Lesson 16-4 Area of Rectangles

Try It

1. What does each square on the grid represent? Find the area of the figure.

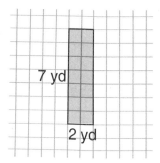

Each square represents _____.

area = _____

2. Find the area of each rectangle.

   a. length = 6 cm, width = 8 cm
   _____

   b. length = 13 in., width = 8 in.
   _____

   c. length = 8 m, width = 9 m
   _____

   d. length = 40 ft, width = 4 ft
   _____

3. Which unit of measure would be most appropriate to measure each area?

   For a and b, write in^2, ft^2, or mi^2.

   a. photograph _____

   b. Antarctica _____

   For c and d, write cm^2, m^2, or km^2.

   c. floor of a gymnasium _____

   d. cat's bed _____

4. Measure the figure. What is the area of the figure? Circle the letter of the correct answer.

   A 18 cm

   B 18 cm^2

   C 20 cm

   D 20 cm^2

5. A bag of grass seed covers 600 ft^2 of land. Can Edwin seed a rectangular area of his yard that measures 14 feet by 40 feet? Explain.

   _____
   _____
   _____

6. What is the total area of the figure? *Hint:* The figure is made of two rectangles.

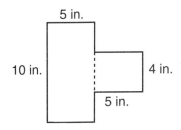

_____

Objective 16.4: Measure the area of rectangular shapes by using appropriate units.

Level D

Lesson 16-5 Volume

Model It — A

Words to Know The **volume** is the number of cubic units needed to fill a solid. **Cubic units** are used to measure volume and tell the number of cubes of a given size that are needed to fill a three-dimensional figure.
A **rectangular solid** is a three-dimensional figure with 6 rectangular faces.

Activity

Build a rectangular solid with these dimensions:

length = 3 cm, width = 2 cm, height = 2 cm

Create the first layer:

6 cubes in the first layer

Complete the figure:

2 layers of cubes in the complete figure

12 cubes in the completed figure

Each cube is 1 cubic centimeter (1 cm³).

The volume of the rectangular solid is 12 cm³.

Practice

Build a rectangular solid. Choose a length, a width, and a height.

length = _____ cm

width = _____ cm

height = _____ cm

Create the first layer.

_____ cubes in the first layer

Complete the figure.

_____ layers of cubes in the complete figure

_____ cubes in the completed figure

The volume of your rectangular solid is

_____

On Your Own

Use 18 cubes to create a rectangular solid. What are the **dimensions** of the solid?

_____

What is the **volume**? _____

Write About It

A rectangular solid has one layer. Ra says that the area of the top of the solid and the volume have the same measure. Do you agree? Explain.

Objective 16.5: Estimate or determine the volume of solid figures by counting the number of cubes that would fill them.

Lesson 16-5 Volume

B Understand It

Example 1

A rectangular solid contains 5 layers of centimeter cubes. The first layer is shown below.
How many cubes are in each layer? 12
How many layers are in the figure? 5
What is the total number of cubes in the figure? 60
What is the volume solid? 60 cm³

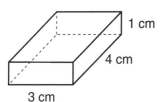

Practice 1

A rectangular solid contains 10 layers of centimeter cubes. The first layer is shown below.
How many cubes are in each layer? _____

How many layers are in the figure? _____

What is the total number of cubes? _____

What is the volume of the solid? _____

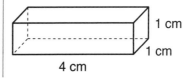

Example 2

The rectangular solid is made from inch cubes.
How many cubes make up this figure? 28
What is the volume of the rectangular solid? 28 in³

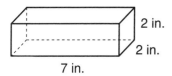

Practice 2

The rectangular solid is made from inch cubes.
How many cubes make up this figure? _____

What is the volume of the solid? _____

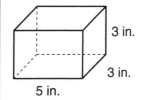

On Your Own

What is the volume of a cube with edges of 6 m?

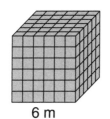

Write About It

Give an example of the units used to measure:

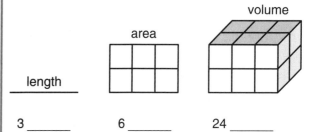

length   area   volume

3 _____   6 _____   24 _____

Objective 16.5: Estimate or determine the volume of solid figures by counting the number of cubes that would fill them.

Lesson 16-5 Volume

Try It

1. The figure is made from centimeter cubes. What is the volume of the figure?

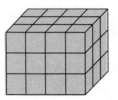

_____

2. The figure is made from centimeter cubes. What is the volume of the figure?

_____

3. What are the dimensions of the rectangular solid in exercise 1?

_____

4. What are the dimensions of the rectangular solid in exercise 2?

_____

5. Find the volume of each rectangular solid. Use unit cubes to construct each figure.

   a. length = 1 yd, width = 6 yd,
      height = 4 yd   _____

   b. length = 3 m, width = 5 m,
      height = 2 m   _____

   c. length = 3 in., width = 2 in.,
      height = 6 in.   _____

6. The first layer of a cube is shown. How many layers will be in the entire cube?

   _____

   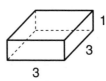

   What is the area of the cube? _____

7. A rectangular box can be filled with 6 layers of cubes. Each layer contains 8 cubes. What is the volume of the box?

   A 14 cubic units   B 24 cubic units

   C 28 cubic units   D 48 cubic units

8. Celina has a box that is 2 feet long, 2 feet wide, and 3 feet high. She claims that 6 square feet can fit inside the box. What mistake did she make?

Objective 16.5: Estimate or determine the volume of solid figures by counting the number of cubes that would fill them.

Topic 16 Measure Geometric Figures

Topic Summary

Choose the correct answer. Explain how you decided.

1. Find the area of the rectangle with a base of 12 inches and a height of 8 inches.

    A 20 inches

    B 40 inches

    C 48 square inches

    D 96 square inches

    _____

    _____

    _____

2. Find the perimeter of a rectangle with a length of 7 m and a width of 6 m.

    A 42 square meters

    B 42 meters

    C 26 meters

    D 13 meters

    _____

    _____

    _____

Objective: Review measuring geometric figures.

Topic 16 — Measure Geometric Figures

Mixed Review

1. Find each product.

a. $7 \times 8 =$ _____

b. $6 \times 9 =$ _____

c. $4 \times 4 =$ _____

d. $5 \times 6 =$ _____

e. $9 \times 8 =$ _____

Volume 4, Lesson 9-5

2. Convert each measurement to the given unit.

a. 20 m = _____ cm

b. 5 ft = _____ in.

c. 800 cm = _____ m

d. 12 yd = _____ ft

e. 24 in. = _____ ft

Volume 5, Lessons 14-1 and 15-2

3. Perform each operation.

a. $46 + 18 =$ _____

b. $46 - 18 =$ _____

Volume 2, Lesson 5-5

4. Which of the following is **not** equal to 30? Circle the letter of the correct answer.

A $900 \div 30$ B $9{,}000 \div 3{,}000$

C $90 \div 3$ D $9{,}000 \div 300$

Volume 4, Lesson 12-1

5. Write each number in expanded notation.

a. 421 _____

b. 610 _____

c. 509 _____

Volume 1, Lesson 2-2

6. A rectangle has a length of 4 cm and a width of 8 cm.

a. What is the perimeter? _____

b. What is the perimeter if the length of each side is halved? _____

Volume 5, Lesson 16-2

Objective: Maintain concepts and skills.

Words to Know/Glossary

A

acute angle — Two rays formed whose angle is less than 90 degrees.

angle — An angle is formed from two rays.

area — The amount of surface inside a 2-dimensional shape. Area is measured in square units, such as square inches (sq in.) or square centimeters (sq cm).

C

capacity — The amount a container holds.

centimeter — A centimeter is a metric unit used to measure length or distance. 100 centimeters equal 1 meter.

cup — A cup is used to measure the amount of a liquid. 1 cup equals 8 fluid ounces.

E

equilateral triangle — Triangle that has all sides the same length and all angles have the same measure.

F

fluid ounce — A fluid ounce is used to measure the amount of a liquid. 8 fluid ounces equal 1 cup.

foot — A foot is used to measure length. 1 foot = 12 inches.

G

gallon — A gallon is used to measure the amount of a liquid. 4 quart equal 1 gallon.

gram — A gram is a metric unit used to measure mass. 1,000 grams equal 1 kilogram.

H

height — The height of a figure is the length of a perpendicular line between the base and the top of the figure.

hexagon — A polygon with six sides.

I

inch — An inch is used to measure length. 12 inches = 1 foot.

isosceles triangle — Triangle that has two sides the same length and two angles with the same measure.

K

kilogram — A kilogram is a metric unit used to measure mass. 1,000 grams equal one kilogram.

kilometer — A millimeter is a metric unit used to measure length or distance. 1 kilometer equals 1,000 meters.

L

liter — A liter is a metric unit used to measure liquid capacity. 1,000 milliliters equal 1 liter.

M

mass — Mass is the amount of matter in an object.

meter — A meter is a metric unit used to measure length or distance. 1,000 meters equal 1 kilometer.

milliliter — A milliliter is a metric unit used to measure liquid capacity. 1,000 milliliters equal 1 liter.

millimeter — A millimeter is a metric unit used to measure length or distance. 1,000 millimeters equal 1 meter.

O

obtuse angle — Two rays formed whose angle is greater than 90 degrees but less than 180 degress.

octagon — A polygon with eight sides.

ounce — An ounce is used to measure weight. 16 ounces equal one pound.

P

parallel lines — Lines that never meet or cross and are always the same distance apart.

pentagon — A polygon with five sides.

perimeter — The distance around all sides of a figure.

perpendicular line — Lines that intersect at right angles.

pint — A pint is used to measure the amount of a liquid. 2 pints equal 1 quart.

polygon — A closed figure made up of 3 or more line segments that meet but do not cross.

pound — A pound is used to measure weight. 16 ounces equal one pound.

Q

quadrilateral — A 4-sided polygon.

quart — A quart is used to measure the amount of a liquid. 4 quarts equal 1 gallon.

R

ray — A line segment with one end.

regular polygon — A polygon with sides all the same length.

right angle — Two rays formed whose angle is exactly 90 degrees.

right triangle — A right triangle has one right angle.

S

sides — The line segments that make up a polygon are called sides.

straight angle — Two rays formed whose angle is exactly 180 degrees. This is called a straight line.

T

ton — A ton is used to measure weight. 2,000 pounds equal one ton.

V

vertex (plural: vertices) — The points at the corners of a geometric shape.

Y

yard — A yard is used to measure length. 1 yard = 3 feet.

Word Bank

Word	My Definition	My Notes

Word	My Definition	My Notes

Index

A
acute angle, 3
angles, 2–4
area, 50–52
    of rectangles, 53–55

C
capacity
    liquid, 23–25
    metric units of, 38–40
centi-, 29
centimeter, 44
cup, 24

D
decagon, 6
drawing angles and lines, 2–4

E
equilateral triangles, 8

F
fluid ounce, 23–24
foot, 17–18

G
gallon, 24
gram, 29–31, 36

H
hept-, 6
heptagon, 6
hex-, 6
hexagon, 5, 48

I
inch, 17–18, 44
isosceles triangle, 8

K
kilo-, 29
kilogram, 31, 36

L
length, 17–19
    geometric, 44–46
    metric units of, 32–34
liquid capacity, 23–25
lines, 2–4
liter, 29–30, 38

M
mass, 35–37
meter, 29–30
metric
    prefixes, 29–31
    units of length, 32–34
    units of mass, 35–37
    units of capacity, 38–40
milli-, 20
milliliter, 38
Mixed Reviews
    13: Basic Geometric Figures, 15
    14: Standard Measurement, 27
    15: Metric Measurement, 42
    16: Measure Geometric Figures, 60

O
obtuse angle, 3
oct-, 6
octagon, 6, 48
ounce, 21, 23

P
parallel angle, 3
parallelogram, 11
pent-, 6
pentagon, 5, 48
perimeter, 47–49
perpendicular angle, 3
pint, 24
polygons, 5–7, 47
    regular, 48
pound, 21
properties
    of quadrilaterals, 11–13
    of triangles, 8–10

Q
quad-, 6
quadrilaterals, 5-6, 11–13
quart, 24

R
ray, 2
rectangles, 11, 53–55
rectangular solid, 56
regular polygons, 48
right angle, 3
right triangle, 9

S
sides, 5
square units, 50, 53
straight angle, 3

T
ton, 21
Topic Summary
    13: Basic Geometric Figures, 14
    14: Standard Measurement, 26
    15: Metric Measurement, 41
    16: Measure Geometric Figures, 59
trapezoid, 12
tri-, 6
triangles, 5, 8–10
    equilateral, 8
    isosceles, 8
    right, 9

U
units
    of length, 17–19
    of liquid capacity, 23–25
    metric units *See* metric
    of weight, 20–22
U.S. customary system of measurement
    length, 17–19
    liquid capacity, 23–25
    weight, 20–22

V
vertex (vertices), 2, 5
volume, 56–58

W
weight, 20–22

Y
yard, 17–18